Water:
The Universal Healer

Guy Proulx

iUniverse, Inc.
Bloomington

Water:
The Universal Healer

Copyright © 2012 by Guy Proulx

All rights reserved. No part of this book may be used or reproduced by any means, graphic, electronic, or mechanical, including photocopying, recording, taping or by any information storage retrieval system without the written permission of the publisher except in the case of brief quotations embodied in critical articles and reviews.

You should not undertake any diet/exercise regimen recommended in this book before consulting your personal physician. Neither the author nor the publisher shall be responsible or liable for any loss or damage allegedly arising as a consequence of your use or application of any information or suggestions contained in this book.

iUniverse books may be ordered through booksellers or by contacting:

*iUniverse
1663 Liberty Drive
Bloomington, IN 47403
www.iuniverse.com
1-800-Authors (1-800-288-4677)*

Because of the dynamic nature of the Internet, any web addresses or links contained in this book may have changed since publication and may no longer be valid. The views expressed in this work are solely those of the author and do not necessarily reflect the views of the publisher, and the publisher hereby disclaims any responsibility for them.

Any people depicted in stock imagery provided by Thinkstock are models, and such images are being used for illustrative purposes only.

Certain stock imagery © Thinkstock.

ISBN: 978-1-4759-3891-3 (sc)
ISBN: 978-1-4759-3890-6 (hc)
ISBN: 978-1-4759-3889-0 (e)

Library of Congress Control Number: 2012913110

Printed in the United States of America

iUniverse rev. date: 8/21/2012

Water:

The Universal Healer

Dedicated to my grandmother Emily, my father Roland, and my mother Theresa who have all left this world. Their collective spirits whisper through my words.

Contents

Acknowledgments		xi
Chapter 1.	My Awakening to Water	1
Chapter 2.	Humankind's Relationship to Water	9
Chapter 3.	Dehydration and Hydration	15
Chapter 4.	Diuretics' Role in Disease	21
Chapter 5.	Your Body Is Talking. Are You Listening?	27
Chapter 6.	The Healing Process of Water	35
Chapter 7.	How Dehydration Negates the Body's Ability to Heal Itself	39
Chapter 8.	The Order of Things	43
Chapter 9.	What Your Organs Have to Say	51
Chapter 10.	Weight Loss and Water	55
Chapter 11.	You and Your Unborn Child—Pregnancy	63
Chapter 12.	Reversing ADHD without Drugs	67
Chapter 13.	Reversing Fibromyalgia Naturally	73
Chapter 14.	Let It Go	83
Bibliography		89

Acknowledgments

Writing this book has made me realize the number of kind souls, family and friends, who walk with me in this life. This project would never have been completed without all their generous help and support. There are too many to mention all of you; you know who you are.

I need to mention a few people who've played a big role in championing this project to life. I would like to thank my editing team. Over the nine years it took to complete this project, I had four editors: Deirdre Lindsay helped with the first version of the book, and I learnt a lot about my writing styles from her. Rolf Pederson, a newspaper editor, contributed his fine work. Sharon Giles is the editor who massaged my words from start to finish. And lastly the editing team at iUniverse publishing—thank you all.

I must also thank the brilliant artist who was able to take a vision from my mind and create a beautiful painting that will help promote the message of my book. If you need to commission an artist, this is your girl. Thanks, Karen Kesteloot, for a great book cover.

Special thanks to Julie and Bryan Tripp, for your skilled photography.

Mike Dionne, Steve Schlotzhauer, Helene Beaudry, Anna's family, Dr Karen Euller, Lori Wilson, and Rob and Sheila Dice are just a few more of those who've supported me in this endeavour. I am forever indebted to all those who made it possible for me to see the truth about water throughout my career.

This book has gone through several revisions over the last nine years, but it was only a manuscript until I befriended Anna Pilon. Anna is my project manager. She came along and said, "We can turn this transcript into a book. No need to be afraid of publishing." She is the reason you are reading this.

Although my whole family was very supportive, I need to give a special thank-you to my brother Pat and his wife, Kelly, for the loan of their cottage, where much of my inspiration came from. It's no coincidence that I was inspired when around water. Thanks, guys, for sharing.

My brother Paul (I have three brothers and four sisters) said to me, "Your acknowledgment is going to be long, but you know who earns the last mention?"

I said, "Who?"

He replied, "Your massage table."

He was right. None of this would have happened without the help of the massage table. I paid all the professionals who helped me for their services in trade with massage therapy. It was upon my graduation from massage therapy college that I began this part of my journey. The sixteen years of research for this book was conducted on my table.

Thank you all. Take care. I care.

Chapter One

My Awakening to Water

No man was ever wise by chance.
—*Seneca*

Of all 5,488 mammals, humans are the only mammal that, after breast milk, drink anything other than water.[1] Everything else you read here will be supporting that one simple, indisputable truth.

I wish to share my awakening to human hydration, and maybe it will help some of you to have your awakening to water as well.

My journey began with my extensive training in the healing art of massage, an ancient healing art that can be traced back through many past civilizations. The part of the world you are from will dictate your view of a massage therapist. In the country I originate from, Canada, two of the provinces regulate massage therapists: Ontario and British Columbia. These provinces have regulating colleges, the same type of regulating bodies that doctors and other

---

1　International Union of Conservation of Nature

health professionals have in order to ensure that the public is safe. This is policed by their guidelines for their members, the provincial Registered Massage Therapist (RMT). At a conference for massage therapists associations, it was said that Ontario and British Columbia massage therapists are the most trained in the world.

I was somewhat shocked; I knew we had United States covered on mere difference in required class time to qualify. Many states only require 600 to 800 hours of training. We require 2,300 hours of training—three times as much—mostly focused on deeper study of human sciences: anatomy and physiology. Pathology is when human physiology goes wrong; it is the study of sickness of the body. Kinesiology is the movement of the body and the forces affecting all joints in the body. Both are also studied in depth.

I did not realize we surpassed Europe's and Asia's qualifications as well. I am making this point not to inflate my image but to clarify my training and qualifications for the unique messages given in this self-help book.

It has been sixteen years since my graduation from massage therapy college, and when I did, my wife at that time, Deborah, gave me the book *Your Body's Many Cries for Water*, by Dr. F. Batmanghelidj. This book influenced the direction of my healing practice right from the beginning. Dr. B., as he refers to himself, has an interesting personal story of how he experienced his own water awakening. As a young medical doctor during the war, he witnessed the healing powers of water while at a German concentration camp. He was the only doctor amongst the prisoners. He began treating headaches, indigestion, intestinal issues, diabetes, depression, and so on with nothing more than clean water—which was all the Nazis would give him.

While in the United States, he wrote *Your Body's Many Cries for Water*, which has helped many of us and woke up holistic healers around the world to the importance of water. In his book, Dr. B. explains this in medical detail. When I read the book, I was fresh out of massage college, and it made so much sense to me. I wondered why, out of all the training I had just received, hydration was treated as a minor mention in my studies. As a treatment-type massage therapist, I work on specific problems, such as whiplash, chronic headaches, neck and low back pain, diabetes, and fibromyalgia. I have always been in troubleshooting mode, looking for a cause rather than a symptom.

Once I had hydration knowledge, I went there first. I asked each client at the beginning of his or her first treatment about his or her drinking habits. Then for at least thirty minutes of the one-hour treatment, I would educate each client on water. They call me "the water man." I have lost more clients to hydration than all other reasons combined. Why? If I was successful in convincing them to just try hydration for three months, their original complaint would no longer be ailing them.

You can well imagine all the people—all with different personalities, cultures, and beliefs—who've ended up on my massage table. Over the years, my clients gave me the opportunities to observe the healing power of water firsthand. Of course, trying to convince my clients to drink water was not always easy. What might work well for one might have no effect on another. I had to be creative in order to hook all these different people on hydration. I thank all my clients for their valuable input into this book.

One year after graduating, I became interested in herbal science. Like most things in my life, I pursued it with the all-or-nothing approach. I then spent four years on research, mainly on North American and Chinese herbs. During my second year I was introduced to iridology, which of course inspired me to become an iridologist.

For those of you who are not familiar with iridology, it is the science of the iris. In approximately 1861, Ignatz von Peczely, a young Hungarian man, was coming back from the fishing hole and found an owl with a broken leg along the trail. He lived on a farm, so he picked the owl up and brought him home. His parents helped him tend to the owl's broken leg. The boy took charge of the bird's care, with joy. Because of the owl's brilliant eyes, he noticed a mark in its eye at six o'clock, on the same side as the injured leg. The owl was around for six months, until he was ready to leave. The mark faded and got smaller every month.

This young boy grew up to be a doctor, and of course, he had never forgotten his friend the owl. When he reflected on the owl's eye, he wondered if it was possible for the eye to record injury in a code form in the fibers of the iris. He began to look in the eyes of patients who had leg injuries, and to his amazement, he found the marks at six o'clock, just like the owl. The leg is represented at six o'clock, the arm is at four o'clock, and the head is between eleven and one o'clock. The right eye represents the right side of the body, and the left, the left side. Dr. von Peczely went on to map the whole skeletal system. A good iridologist can find years of growth and development recorded in your iris. It has advanced so much that while I lived in Britain I was able to attend a postgraduate iridology workshop to further my research.

Two European iridologists have learnt how to read powerful information about personality and past events in the iris. They are all recorded in the eye. To give you an example: I looked into a patient's eyes and said, "Something happened to you when you were eight, and you got over it by the time you where eleven. Do you remember what it was?"

The patient, Janet, replied, "No, nothing happened to me when I was eight."

So I said, "It is in your mother eye as opposed to the father eye; it could be an influence from your mother."

She said, "Well, my mother died when I was eight." I asked what happened at eleven, and Janet said, "My dad got married, to the same woman he is still with today."

Studying iridology was a natural progression. It is a powerful tool for troubleshooting. It's relatively new to the Americas but much more widely known and practiced in Europe. Many European medical doctors train in iridology.

Iridology helped me to find out what herbs are needed to assist the body regain balance. Everything can be looked up in the eye. We can see if you have parasites, whether or not you are assimilating your food, if you have cancers or beginnings of cancers, what the pH of your body is, and which ones are your weak organs that need and want support. Past injuries and events are all recorded.

While I was promoting a line of herbal products, I lectured a lot—I think mainly because I like to hear myself speak. I used that platform to spread my hydration message as well. I would always start my herbal speeches with a twenty-minute talk about water. To me, it was only logical; the audience came to hear me speak on holistic medicine. Why not push the most powerful one? I impressed upon them that the herbs they were about to try would work most efficiently with clean water; putting medicine into a hydrated body has a much more potent effect than if the herbs enter a dried-up body. Eventually, I was asked permission to film my lecture so it could be played for other groups throughout Canada and the US. At that point, I knew that hydration was a thought-provoking concept for my audiences.

"Could many of my ailments be corrected by simply changing my drinking habits?" The answer is yes.

In 2003, I began writing the biggest essay I have ever written. I finished it in 2005, and I found that publishing it was harder than writing it. Frustrated, I gave up the idea of having my book published.

I was on fire during those years, speaking about water to whoever would listen. In 2004 and 2005, I was the official massage therapist for the Guelph Storm hockey team. Hockey is a serious game here. The Storm players are in the Ontario Hockey League; this is the last step before making it to the professional league, the NHL. These young men do not reach this level of hockey without being extremely coachable. Let me give you an example.

At the beginning of a season, one of the Storm players was at my clinic, escorted by his mother. Being the typical hockey mom, she was frantic in her explanation of what had happened to her son, Todd. Todd had pulled a groin muscle during training, three days before the season tryouts. Even before treating him, I knew which muscles were in spasm. At the end of the first treatment, I showed him the proper stretch to lengthen the muscle in spasm. Then I said to him, "I want you to do this stretch fifteen times a day," and the only response I received from him was "Okay." I told her I would treat him today, bring him back Friday and Saturday, and we would see. By the third visit, Todd was ready for the tryouts.

Just before he left, he asked me, "How long do I have to do this stretch fifteen times a day?"

Shocked, I looked at him and said, "Did you really do the stretch fifteen times a day?"

Todd's response was simply, "Well, you told me to."

I replied, "I said fifteen in hopes you would do it four or five times a day."

I realized then how open these young minds were to instruction. Most of them are fifteen years of age or older and have always dreamed of making it to the NHL. Todd made me realize that I need to be careful what I recommend; the players view me like they view the coach and the trainer—just another part of their ticket to the NHL. I made sure to pass my hydration knowledge on to these players; the more assertive and passionate I was with my delivery, the more impact I would have. I lectured for ninety minutes about water to the entire team, coaches, and trainers.

During my second year with the team, they asked me to lecture at the OHL annual conference in front of all the teams' trainers. I had an hour, and they wanted me to lecture about the mid-region of the body. At the time, I was just finishing my book, and I asked if I could speak on hydration for half an hour. They agreed, and the response I received was empowering. It sparked so many questions that they had to cut the question-and-answer session short so I would have time for the main lecture.

For the next six years, I trained in nutrition at deeper levels and less-traditional nutritional approaches to using food as medicine. A great book on this topic is *The Cure*, by Dr. Timothy Brantley.

I also explored energy work as another tool for my medicine bag. I am a Reiki master, and I studied shaman healing for two years. Energy is present in everything around you right now. Do not ignore energy when troubleshooting a problem.

One day in 2011, a client who coached a high school girls' soccer team commented to me, "I wish I could get the girls on the team to see the value of hydrating." I offered to speak to them after a practice, and he had me there that Sunday. As he gathered the girls after practice and was about to introduce me, one of the parents approached from the parking lot and asked, "Isn't the practice over at eleven thirty?" I invited the parent to listen, stating that it might be of interest to him as well.

These girls were sixteen and seventeen, so I tailored my talk to my audience. I explained the consequences of not hydrating; they could break out in acne—no sweet sixteen wants that—they might gain weight, and their athletic performance would suffer. The entire talk lasted for only fifteen minutes. After I was done speaking, a parent asked, "Do you know the trainer for the Guelph Storm?"

I replied, "If it is the same one who was there seven years ago, yes, I do. Why do you ask?"

He said, "Because I've heard him giving a hydration talk many times. Not as good as you, though."

It had been some time since I had lectured publicly on hydration, and his comment touched me deeply. I was amazed that this trainer—who is responsible in part to help mould these young players to their top performance—still uses my speech as the basis for his talks, seven years later. This trainer had only heard my lecture three times. I thought about the impact I had had on him and all the lasting good he had done for these players. It was humbling. I am sure other trainers around the province also continue to preach the benefits of water to their players. It was then I knew I had to publish this message. Hydration is one of the secrets that help us to dominate the game of hockey.

I am an avid believer that everything happens exactly the way it is supposed to. It is the way we react to what happens that ultimately shapes our journey.

Chapter Two
Humankind's Relationship to Water

What lies behind us and what lies before us are small matters compared to what lies within us.
—*Ralph Waldo Emerson*

Water and Our Planet

Earth consists of more water than land mass. Water gives life to the whole planet and every living being on it. Consider these three amazing facts about water:

- There is not one thing on Earth that water cannot break down. Even diamonds, in time, are no match for water.
- There is nothing that water will not clean or dilute.

Water is the universal solvent.

- There is absolutely no living organism on the planet that does not depend totally on water for its healing powers and life-giving nourishment to survive.

After learning all about the life-giving and life-sustaining properties of water, I have come to the following conclusion:

To intentionally deprive your body of water is tantamount to embarking on a slow, insidious, and premature death.

Consider what the results would be if one-third of our planet's water was suddenly removed (equivalent to the volume of the Atlantic Ocean).

- Millions of people would die.
- Most marine, wildlife, and plant life would perish.
- Disease would flourish.
- Hunger would be the biggest killer.
- Weather would drastically change.
- A massive desert would form.

The same domino effect of destruction and ultimate ruin also results from depriving your body of clean water. Decades of dehydration stress on the body is a contributing cause to headaches, back and neck pain, cancers, fibromyalgia, diabetes, heart problems, intestinal disorders, premature aging, obesity, and other ailments.

A Quick, Important Biology Lesson

Do you remember studying the Krebs cycle in biology class? The Krebs cycle occurs in all cells in your body. This cycle is important because it produces adenosine triphosphate (ATP), the energy of the body. Nothing happens in the body without ATP's involvement.

ATP is the power used in every biological function in your body. Cells use ATP to transfer energy, not to store it. ATP is an energy-carrying molecule, which transfers relatively small amounts of energy as a fuel molecule to the points in the cell that require energy. The Krebs cycle requires one free hydrogen molecule to produce twenty-six of these irreplaceable molecules.

So, what is the easiest way for the cell to access a free hydrogen molecule? The answer is water. The body uses the hydrogen in water (H_2O) and glucose (sugar) to produce these powerful energy molecules. In turn, this process allows the oxygen molecule to carry precious oxygen to all the parts of the body.

Without clean water, the very beginning of life-healing energy in your body is drastically reduced!

The human body is a marvel that, to this day, man does not totally understand and may never totally understand.

The Body—A Complex Marvel

Here are a few facts we do know about the human body:

- A human blood cell lives for approximately 120 days and then is recycled.
- The liver separates nonessential properties from the dead cell and extracts iron.
- Iron carries oxygen in the blood and can be recycled several times.
- The liver sends the iron back to the marrow in the bones, where new blood cells are being manufactured.

Our bodies are constantly producing new life within. In fact, we get a new liver approximately every five months (not curing a disease of the liver, but just through the replacement of dead cells). We also replenish our skin approximately every four weeks. This life-renewing ability is precisely why the human body can be submitted to such unthinkable conditions such as wartime atrocities, torture, assault, and cancers—and survive to talk about it.

The human body is like a slow-flowing river, constantly changing and adapting to its environment. That's why you can be severely dehydrated when you are young, and it will go undetected until you reach thirty-five to forty-five years of age, when you have a dehydration awakening.

Water's Life-Sustaining Properties

Let's illustrate the life-sustaining properties of water using a simple houseplant. A relatively simple organism compared to mammals, a houseplant may completely fall over if deprived of water. If there is any life left, all you need do to revive it is provide it with life-giving water. Roughly an hour later, the houseplant will be standing up and only because of water.

The soil may contain the minerals, but the plant cannot synthesize minerals without water. No ATP means no energy and poor results. So instead, the plant's roots will react by withholding precious water from the part of the plant above the ground. The plant will store the water in the roots in order to ensure its own survival.

If humans had such a limited ability to cope with dehydration, roughly 85 percent of the population would be lying on the ground, waiting for a kind soul to pour water down their throats so they could stand up!

Look at water as the most important food your body requires. You can also see it as the natural medicine it is—it provides freedom from certain ailments and acts a cleanser or as a means of restoring balance in the body. You will learn to love consuming water to quench your body's appetite.

Even if you don't perceive water as the most important food, your body certainly does!

If you know people who have a hard time keeping excess weight off or losing those last ten pounds, there is a good chance that they are not drinking enough water. These people are likely dieting in a state of dehydration or starvation.

It is no coincidence that all successful weight-loss programs first and foremost insist that you drink ten to twelve glasses of water a day. If you do not consume enough water, you will not be able to achieve your target weight-loss goal. The reason is simple: deprive yourself of water, and you deprive yourself of the most powerful food you need. Your body will go into starvation mode, and we all know how starvation diets work. You lose weight at first, and then the body goes into a state of crisis because it is starving.

In starvation mode, the body takes any food you ingest and makes you tired so you don't waste energy, and then it turns that food energy into fat—just like the plant, the body is storing food. In effect, the body is preparing to cope with its starvation and dehydration crisis. Many dehydrated people are misdiagnosed as having chronic fatigue syndrome when they are just thirsty. When we remember that we are part of nature instead of seeing ourselves as better, separate from, or above it, the logical conclusion is that we are functioning in an unnatural state if we are not drinking the proper amount of water each day.

Chapter Three

Dehydration and Hydration

*Courage does not always roar.
Sometimes, courage is that quiet voice
at the end of the day saying, "I will try
again tomorrow."*
—Mary Anne Radmacher

How do you know if you are operating in a hydrated mode or a dehydrated mode? Your body is either in one state or the other, and it's important to look at it from a human physiological and pathological perspective.

Most of us are familiar with the magic numbers of ten to twelve glasses (approximately three litres) of purified water each day. Realistically, if we calculate our diuretic fluid intake (fluids that stimulate water loss) and refuse to give them up, most of us need more like twenty to twenty-four glasses of purified water per day, just to break even. On average, that amount is required for the body to attain proper hydration maintenance at a systemic level (any and all systems, including the nervous system, digestive system, immune system, and circulatory systems). Keep in mind that our systems want to function in harmony.

Water Is Water—Or Is It?

Whatever type of fluid you consume, the kidneys and liver will turn it into distilled water before the body can use it. Distilled water is pure H_2O.

Water is the prime building component of new tissue or cells.

Much of the tap water you drink is used in repair work. It is needed to replace and clean cells killed by chlorine and other toxic substances. You need to consume additional water if you drink anything other than distilled water. Clustered water is also very good[2]. This type of water allows your kidneys and liver to extract the minerals from the water. Clustered water contains certain minerals that are in a form our bodies can assimilate; unlike spring water, we can't assimilate those minerals. Clustered water is mainly derived from melting glaciers.

If you have a dog or cat, you might have noticed the following behaviour. When you know your animal is thirsty, you fill its dish with fresh tap water. Initially, the animal will go to the dish and put its nose above it and walk away from it without drinking. In some cases, the animal will go outside and drink from a puddle of dirty, stagnant water or even drink out of the toilet. A couple of hours later, it will come back to drink from the dish. The reason for this behaviour is simply that dogs' and cats' sense of smell is significantly more refined than ours. When your pet inspects the dish before drinking the tap water, the blast of chlorine evaporating from it will be overwhelming. Your pet's own survival instinct induces it to move away from the dish even though the animal is thirsty. When it returns a few hours later, a good portion of the chlorine will have dissipated.

---

[2] Brantley, Timothy. *The Cure*. New Jersey: John Wiley & Sons. 2007.

Man has learned many powerful health remedies from animals. If you use a relatively simple form of water filtration for yourself, such as a portable countertop container, you should leave the top off and let the water sit at room temperature for a while. This will give the chlorine a chance to evaporate before you consume it.

Distilled water does not require any further processing by the body before it is transferred into pure energy. Distilled water is the only water that helps the body rid itself of heavy-metal toxins that accumulate in joints and muscles.[3] Thus, consuming distilled water removes dehydration stress systemically throughout the body.

If you take in three litres of tap water, a good part of it will be wasted in repairing the damage caused by chlorine, lead, and other toxins present in your city water. Consider what one drop of bleach does to a brand-new pair of jeans. Bleach is the same chemical composition as chlorine; imagine how many live cells it can destroy in your body when you consume it. The chlorine may be at what's considered acceptable levels, but you should be the one to decide how much poison is okay!

Relieving your body of any stress is always a move in a healing direction.

The next-best form of purified water is through the process of reverse osmosis. One downfall to reverse osmosis is that while it is clean, it is also void of positive and negative ions. For that reason greenhouses and nurseries typically do not use it.

Prior to researching this topic, I thought spring water must be the best form of drinking water available, mainly because of its high mineral content. However, the minerals in spring water are in a crude form that the body cannot completely assimilate. Nonetheless, spring

---

[3] Walker N. W. *The Natural Way to Vibrant Health*. Caroline House, 1976.

water is still a better choice than tap water. Plants drink spring water and synthesize the minerals in a form that our bodies can successfully process and assimilate. Therefore, we can best take in minerals by eating raw foods and herbs.

If the minerals in the water we drink are ionized or clustered, some (like calcium and iron) can be assimilated by the body.[4] The most popular source of ionized water is that which comes directly from glaciers.

Diuretics—The Two-for-One Concept

Does this mean that all you need do to attain better health is drink two and a half to three litres of distilled water daily? For approximately 25 percent of the population, the answer is yes. Over time, you will have restored balance to your body. The other 75 percent or so of people who drink that much will likely feel healthier simply because they are consuming more precious water, but diuretics must be taken into consideration.

For every diuretic taken in, you lose approximately two glasses' worth of water. Coffee, black, green, and iced teas, soft drinks or soda, chocolate, and alcohol are all diuretics. Other than pure water, only unsweetened herbal teas count toward your daily water intake.

On an average day, for example, you might drink two cups of coffee or tea, perhaps a soft drink at lunch, and one glass of wine with dinner. On that day, you will have consumed four diuretic drinks, but your body will have lost the equivalent of eight glasses of water. Since your optimum daily water intake should be twelve glasses, you would need to drink twenty glasses just to break even that day.

Reflecting on your past from this point of view, how many times in your life have you truly existed in a hydrated state?

---

4   Brantley, Timothy. *The Cure*. New Jersey: John Wiley & Sons. 2007.

To this point, hydrating yourself may seem a simple, straightforward procedure. Factoring in diuretics, you might now have to come up with a completely new health strategy. Breaking harmful old habits is hard work, but to commit to healing your body you must drink the recommended 2.5 to 3 litres of water each day.

If you are a baby boomer (roughly forty-six to sixty-five years old in 2012), your farm-raised grandparents would have considered it a treat to have coffee with Sunday dinner. They drank water primarily. Seasonally, they would have squeezed fresh fruit to make juice. Unlike us, our grandparents did not drink diuretics on a daily basis, nor did they suffer from the alarmingly high incidence of diseases we now face, diseases such as cancers, heart disease, diabetes, chronic fatigue, fibromyalgia, and intestinal problems.

Chapter Four

Diuretics' Role in Disease

> *The most beautiful thing we can experience is the mysterious. It is the source of all true art and all science. He to whom this emotion is a stranger, who can no longer pause to wonder and stand rapt in awe, is as good as dead: his eyes are closed.*
> —*Albert Einstein*

If I asked you to give your pet only coffee, tea, soda, and alcohol for one month with no water, what would you say? I'm sure you wouldn't entertain that horrible thought for more than an instant; you know it would be inhumane and would ultimately harm your pet. But you are also an animal, and if this is what you habitually drink month after month, aren't you also harming yourself?

A Cellular Perspective

How do diuretics affect the body? At a microscopic level, everything, including water, needs an invitation to enter a cell. Water has to follow sodium (salt) through a sodium gate in order to enter a cell.

There are various kinds of receptors and channels in the cell wall, corresponding to calcium, iron, magnesium, and vitamins—everything needed to allow passage. Receptors are a certain shape and will only permit a twin shape to pass through into the cell.[5]

One of the most harmful effects of toxins is that they can block the entrance of nutrients to the multitudes of receptors and channels. That means the cell isn't able to survive in a healthy and productive manner. A diuretic such as caffeine or alcohol does not require an invitation to enter a cell. Diuretics bypass the receptors by breaking through the cell wall, thereby killing the cell. This is how diuretics cause you to lose twice as much fluid as you take in; the water is consumed to clean up, recreate, and replace the dead cells.

Continual Growth—The Body's Defence against Aging

The body is continually adjusting to its environment. Remember the days when a full forearm cast was applied to a sprained wrist? When it was removed seven to ten days later, the arm was visibly smaller and weaker. After just days of disuse, the body atrophied (broke down) the muscle tissue and used the protein for other parts of the body. The old adage "use it or lose it" is applicable to every living organism.

Your body reacts the same way to dehydration. When you are severely dehydrated, your body focuses on one thing: maintaining a safe, systemic operating level throughout. Your body will take drastic

---

5   Widmaier, Eric P., Hershel Raff, and Kevin T. Strang. *Vander, Sherman & Luciano's Human Physiology: The Mechanisms of Body Function.* McGraw-Hill. 2003.

measures to conserve precious energy. This will be explained in greater detail in the next chapter.

In an attempt to save large amounts of energy and minimize body movement, you may become extremely fatigued (chronic fatigue syndrome). As stated above, when you consume diuretics, you lose large amounts of irreplaceable fuel or energy. Your body's immune system, which polices and orchestrates repair throughout your body, is severely affected by this abuse. When you are improperly hydrated, the immune system will spend most of its time finding, fighting, and repairing damage caused by dehydration stress.

The human body is constantly searching for ways to conserve energy. The more dehydrated you are, the more frantically your body searches for any way to restore the optimum balance.

A Revelation (Not a Test)

Using what you have learned so far, your first step to determine if you are properly hydrated is to review your water and diuretic intake. If you discover that you are operating in a severely dehydrated state, you will need to establish how many years you have been living in this damaging state.

You will also need to determine where you rate on the hydration scale. To do this, document how much pure water you consume in any given day. (Remember that only herbal teas can be included as water intake.) For every diuretic taken in, you will need to subtract about two glasses of water.

Don't be discouraged if you only come up with two glasses of water per day or if you are in the negative. You are not alone. Lucky for you, the situation is easily corrected. Complete the following exercise, and be very honest with yourself. This is not intended as a test but rather as a personal revelation.

Hydrated or Dehydrated?

- Total your average daily intake of water. Be painfully honest about everything else you drink.
- Total the diuretic fluids or foods you consume.

Keep in mind that drinking ten to twelve glasses of purified water per day is the ideal.

Example:

If you consume two cups of coffee a day, one soft drink, and two glasses of wine with dinner, you have consumed a total of five diuretics. That means you have lost ten glasses of water. If you consumed six glasses of distilled water on that same day, you are at negative four. Since you need approximately eleven glasses per day, you would still need to drink another fifteen glasses of water, for a total of twenty-one glasses of water, to offset the effects of the diuretics. A remedial plan is to give up some or all of the diuretics in your diet.

Tip:

Purchase a one-liter water bottle and fill it with distilled water at the beginning of your day. Put two elastic bands around it and take this bottle everywhere with you (in the car, at work, by your bed at night, etc.). Once it is empty, fill it up again and then remove an elastic band. At the end of the day, you will know how much water you have consumed.

Now sit down and make a list of every little thing that bothers you in every facet of your life. Nothing should be spared (e.g., your husband, your children, someone else's children, coworkers who irritate or anger you, road rage, lack of patience, your work, your boss, headaches, back/neck pain, acid indigestion, colon problems, dizziness, fatigue, lack of incentive to exercise). Be creative and have fun. Then put the list away, go to your planner or calendar, and make a note to reread the list three months from now. If you stay hydrated for that length of time, check your list. I predict that 80 percent or more of that list will not be as bothersome in your everyday life.

I challenge you to prove me wrong!

Chapter Five

Your Body Is Talking. Are You Listening?

*I'm an idealist. I don't know where
I'm going,
but I'm on my way.*
—Carl Sandburg

At this point, you must be thinking, *If I am causing this cascade of destruction in my body, surely I would have visible and physical signs of it?* You do, but you may be unaware of your body's physiological response to dehydration.

In the heat of the summer, the signs of dehydration are everywhere. Grasses cease to be plump and green. Instead, they turn yellow. In order to save the entire plant from destruction, the plants stop trying to maintain the lush green leaf. The leaf is deprived of water and nutrients, but the plant stores adequate food so that it is able to recover once rain arrives. During this dry period, forest fires become frequent occurrences, crops suffer and are often severely damaged or lost altogether, and trees seem to slouch from lack of moisture. Certain insects, such as chinch bugs, ants, and grasshoppers, thrive and multiply in conditions of drought. Most other insects and animals aggressively avoid the heat so they can conserve precious energy. Some animals begin to emerge only at night.

Maybe your body is giving off signals that you are dehydrated. Are you hungry between meals or hungry one hour or so after a meal? This is your body calling for fuel in the form of water. After a large meal, the body asks for water to help digest the food eaten. It is recommended you allow thirty minutes after meals before drinking fluids, allowing the digestive enzymes in the food to work before diluting them.[6] When you are not consuming enough water, the body will encourage systems in the body to give up some of their water. The body defends itself by reducing muscle movement so the water can be directed to the stomach, liver, and colon.

We hit the couch after a Christmas or Easter dinner in order to conserve the energy needed for digestion. A large carbohydrate intake also contributes to the obvious energy loss; the body takes this excess sugar and manufactures fat as stored energy. The body requires more energy to produce fat than to break it down, so when the thirst signal is activated and you reach for a diuretic or more food, your body will shut you down. You will sit immobilized on the couch. You are too fatigued. Inevitably you will fall asleep earlier than you had planned. If you are often hungry between meals, quench your hunger with water. It is calorie-free, and you won't add a single pound by drinking it.

You may complain, "I am just so tired when I get home from work!" Your commute may involve driving an hour or so to work on a busy highway at the beginning and end of your day. When you arrive at

---

[6] Batmangheldij, F. *Your Body's Many Cries for Water.* Falls Church Global Health Solutions, Inc., 2008; and Brantley, Timothy. *The Cure.* New Jersey: John Wiley & Sons, 2007.

work, there is a day and a half of work sitting on your desk. There are countless other scenarios that could cause you to bear extra, unhealthy stress. To name a few:

- disagreements with your supervisor
- quarrelling with your child
- money problems
- car repairs

How many of these situations are part of your life? We are juggling a lot, and if you are dehydrated, you have unnecessarily added more stress to the mix. But if you listen to your body and drink water when it asks you to do so, you will feel the difference in your energy and stress levels.

Most people will experience a major energy drain forty to sixty minutes after dinner. If you were dehydrated to begin with, you will be operating close to empty. No wonder you're tired. This critically low level of energy that you have been struggling with all day becomes physically obvious. Dehydration is the number-one reason for a lack of energy at the end of the workday, in my opinion.

Water also plays an important role in managing weight loss. Proper and permanent weight loss also includes making several lifestyle adjustments. To achieve pain-free weight loss, you require some cardiovascular exercise, an intake of essential fatty acids (EFAs) and fibre, a parasite cleanse, higher protein consumption, and lower carbohydrate intake. However, if you did all of these things but did not drink the proper amount of distilled water, you would still not meet your weight-loss goal. By putting your body on a starvation diet, you simply might not be able to burn off those last ten pounds.

I Am Never Thirsty!

You are not a camel. Do not be misguided; the thirst signals are there for you. Did you know that the last sign of complete dehydration is a dry mouth? I once believed a dry mouth would be the first indicator. When you experience a dry mouth and choose something other than water to quench your thirst, your mouth will remain dry.

A hangover is nothing more than a dehydration headache. Remember that alcohol, like caffeine, attacks and destroys cell walls. Your body is in desperate need of water to repair that damage. Of course, you can acquire a headache without the help of alcohol. Raised levels of stress and an excessive caffeine intake can lead to instability in almost any bodily system. Toxins, such as alcohol, coffee, and cigarettes, are major contributors. So next time you experience a headache or a hangover, drink a liter and a half (six glasses) of distilled water, even if you cannot keep it all down. Ninety-five percent of the time, you will feel normal in one hour.

Dry Skin, Sore Muscles and Joints

The brain targets the skin as the first organ from which to retrieve water. Eventually, the muscles, joints, and lymphatic system are also targeted. Your skin dries up in the winter often from the inside out. In winter, we are inclined to reach for warm drinks, such as coffee, tea, and hot chocolate, to combat the cold. A tall, cold glass of water is probably the furthest thing from your mind at this time of year, but the severe contrast between the temperature outside and the temperature inside causes you to burn more water in order to maintain an ideal core temperature.

In addition to this, dehydration means you are unable to properly cleanse the muscle tissue of lactic acid. Lactic acid is produced when we stress our muscles. Lactic acid sits on the muscles when the lymphatic system, which rinses and cleans the tissues, dries up.

When you take your first bike ride of the year and walk funny the next day, this is lactic acid pain. Muscle pain following a physical workout is due primarily to excessive micro-tearing of muscle tissues, coupled with an inadequate repair mechanism. This is a result of dehydration. When I worked with dehydrated hockey players, they experienced injuries or high levels of tearing because of the level of physical activity in hockey. By midseason, many of the minor injuries disappeared simply because the players were drinking enough clean water.

Your joints are all surrounded by synovial sacks. Synovial fluid is contained in these sacks. Synovial fluid lubricates the cartilage in and around the bones and joints. This prevents bone-on-bone friction and excessive cartilage wear.

People with arthritis will likely have inadequate synovial fluid levels and possibly damaged synovial sacks as well. Water is required in the body's production of synovial fluid. Dehydration will eventually result in chronic muscle and joint pain.

Why Am I Plagued with Constipation, Diarrhea, Heartburn, and Ulcers?

Most of these symptoms can be eradicated simply by adjusting the body's moisture levels. Once the brain has effectively drained as much water it can from the skin, muscles, and lymphatic system, it turns its attention to the colon. The kidneys will receive instructions from the brain to extract water from the colon, clean it, and send it elsewhere in the body. In an extended period of dehydration, the colon and the entire digestive track will be severely compromised. Most cases of constipation and diarrhea are caused by a deficiency of water in the colon. That is why the same remedy—an increase in fibre and water intake—will usually take care of both.

Constipation

Constipation happens in the large intestine. Unlike the small intestine, which is responsible for the digestion of food, the large intestine moves waste left over from the digestion process. When the brain continually signals the body to take water from the colon for use elsewhere in the body, the fecal matter in the colon dries up. The dried-up mass can create a blockage. Over time, you develop what is called a lazy colon. Remember, if you don't use it, you lose it. When it is not getting enough water, the colon does not undergo peristalsis (a pumping action, like a moving snake), which is the muscle movement that pushes fecal matter through it. As it weakens, the colon continues to plug. This happens over time, and layers and layers of toxic fecal matter can become imbedded against the inner intestinal wall and dry to a semisolid mass. From there, toxins may seep through the intestinal walls and, in time, enter the bloodstream. This may result in

- headaches;
- fibromyalgia;
- attention deficit hyperactivity disorder (ADHD);
- cancers;
- heart-related conditions;
- digestive problems.

A poorly functioning colon is cited as one of the underlying conditions in more than 60 percent of all cancers. Related colon problems probably afflicted these cancer patients for most of their lives. This is an organ you want to pay special attention to. A form of natural fibre supplement that has a mixture of soluble and insoluble fibre and four litres of water a day for two weeks will, in most cases, have you feeling a lot better, but do not stop taking fibre and water when those two weeks yield the desired results. The water and fibre will nourish, scrape, and cleanse the inner colon walls. During a constipation/

dehydration crisis, organs that normally dump waste products into the colon are denied access. They will not be cleansed properly. When the colon is allowed to operate efficiently, the entire gut will be efficient. Other organs will then be able to send their toxins to the colon for disposal.

If the colon is the major toxin-dispensing organ, should we not encourage it to keep moving?

Diarrhea

The same primary developments that occur in constipation will occur in diarrhea. That is, there will be a compacting of thick, toxic fecal matter in the walls of the colon. Constipation will result from partial blockages. However, with diarrhea, the body will form a small passageway through the compacted fecal matter, about the diameter of your baby finger, to successfully move liquid past the blockage in your colon. Again, ingesting fibre and water will help regulate your system and repair the damage to your colon, thus restoring a healthy balance.

Diarrhea and constipation will contribute to conditions such as

- headaches;
- indigestion problems;
- heartburn;
- ulcers;
- diverticulitis;
- Crohn's disease;
- colitis;
- fatigue.

The body often informs you when it needs water. You may have simply never associated dehydration with these common symptoms.

Chapter Six
The Healing Process of Water

A wise man can learn more from a foolish question than a fool can learn from a wise answer.
—*Albert Einstein*

Before you can begin to understand the magical healing powers of water, you must first understand the damage dehydration does to the human body. If you have been dehydrated for years, then you have been conditioning your body to operate quite like the Sahara desert. This might sound funny, but it's true. When the body needs water, it will work within its predetermined parameters to ensure its survival.

All organs are instructed to operate on less water. This means the heart is being told to slow down so as not to waste precious water. It is under stress to make do with less energy, but nothing may have been done to cut back on its workload. In fact, a lack of water will increase the body's workload! This will affect the skin, bones, muscles, lymphatic system, lungs, liver, spleen, intestines, and so on.

People in this state will often say they are stressed out. Because of the deprivation of power and resources, a cascade of stresses to most parts of the body will be triggered. Hydrating is a simple way of alleviating massive amounts of internal stress.

Now, if dehydration has been chronic, your body will be attuned to your watering habits. It will not have the capacity to absorb two and a half to three litres of water a day when you begin to hydrate. The muscles and organs will still be in starvation mode. High and consistent daily consumption of water will be needed to switch off the established starvation mode.

When you are hydrated, it is natural to have four to six healthy urinations a day. This is not only good for you, it is necessary. You will notice this as your body is hydrating. The body will simultaneously signal your pituitary gland to increase growth hormone production for your bladder and urinary tract. This will help strengthen your bladder. This organ will not yet be capable of processing that much fluid. While the bladder is adjusting, you will receive signals to urinate. Resist those signals, but not to the point of pain. This will send a strengthening signal to the bladder. This is the same advice given by some physicians to children who wet their beds.

If you can stay on this hydration track for four to six weeks while maintaining your vitamin and mineral intake, you will be amazed at how your health will improve. All you need to do is drink the proper amount of clean water. Just a note of caution: that extra water might cause you to lose some electrolytes. Consider taking a good quality multivitamin tablet or local bee pollen.

If we breathe, drink, or eat, we are ingesting toxins, which must be efficiently expelled. Water is the most powerful tool there is in breaking down, transporting, and removing toxins from the body. Just bring back the balance and experience the healing.

Whenever you feel tired and listless, it probably means you are down a liter or two of water. Try drinking one and a half litres of water at once, wait an hour, and then feel your energy change. When there is ample water, the immune system purrs like a cat. As long as it is purring, you are healing.

It is said that 80 percent of all healing occurs during the first two or three hours of deep REM sleep. So, if you are waking up in the morning and still feeling tired, one of three things may be happening:

- You are going to sleep depleted of water energy.
- You are attempting to enter a state of major healing without the needed energy. (new bullet) Or your sleeping position is negating it.

The cat is not purring and you are not going into deep sleep. Your flight-or-fight or sympathetic nervous system is on full power. In that state, the flight-or-fight reflex overrides the parasympathetic, or "housekeeping," nervous system, which is in charge of maintaining the body's balances in temperature, blood flow and distribution, and nervous, digestive, and lymphatic systems. This explains many physiological states of what is commonly referred to as nervous breakdown.

Let's look, for example, at someone who has taken on enormous financial debt, commutes to the core of the city and back daily for three hours on a good day, doesn't get along with his or her boss, is in a stressful job, and is having marital problems.

This person may wake each morning tired and possibly vomiting bile, never feeling totally rested and revived. He or she must reduce the stress he or she has been enduring to avoid a nervous breakdown. This person might not have experienced deep, parasympathetic sleep in years. When a person has a nervous breakdown, he or she will often be sedated to allow the parasympathetic nervous system to begin the healing process. You need to be mindful that your fight-or-flight system is turned on.

In addition, you may add some activities to your life to facilitate nervous system healing. Meditation, massage, yoga, Pilates, Tai Chi, and other forms of exercise, including sex, will all benefit the parasympathetic nervous system. This is how it works:

- Reduce stress.
- Eat healthily.
- Add proper amounts of clean filtered water.
- Be patient and consistent.

And you will rebalance yourself!

Chapter Seven

How Dehydration Negates the Body's Ability to Heal Itself

*If you're not riding the wave of change,
you will find yourself beneath it.*
—Author Unknown

Imagine an extended heat wave in a desert. That is truly a dry season. The hot, desert sun will bake the life out of the soils. The nights will be brutally cold, the days blisteringly hot. Earth is a master of repairing and protecting itself. The desert soil will be protected and act as an insulator to protect the life stored within the root systems and seeds below.

Imagine how this plays out. The desert is so silent it appears to be lifeless, but just days after the rainy season returns, it teems with plant life. Flowers and little shrubs have popped up everywhere. Birds, mammals, and insects congregate to celebrate the sudden presence of water.

Following monsoon rains, the plants' roots will descend deep into the ground to store energy in the deepest part of the root system. The plant hibernates to avoid death from the severe conditions. Do you think water alone is responsible for evoking this radical change? Of course!

The resilience of nature is unrivalled.

Let us compare and contrast the desert plant and its adaptive reactions to the way some of us treat our bodies. The skin muscles and skeletal frame are akin to that part of the desert plant existing above the ground. The nervous system, heart, lungs, and other organs are analogous to plant roots.

If you truly are dehydrated, it is safe to say you are impairing the health and the ability of your body to perform basic functions. These behaviours dictate your past, your present, and your future level of health. After you have made a few minor adjustments in your life and increased your water intake, you will reap huge rewards. You do not know how good you should feel because the negative change has been slow and relentless.

You must next ask yourself how long you have been dehydrated.

Be careful of the ruts you choose in life, because you will be in them for a long time.

Your body can become adept at operating on low to no free water. The brain will not function naturally without its valuable water energy. So the brain will first ensure there is enough water for itself. The blood-brain barrier very carefully selects what gets through to be consumed. The function of this filter is to block any alien organisms or toxins from passing through. Unlike other parts of the body, the brain will not break down muscle and turn fat into energy. The brain feeds only on the purest form of food, glucose.

The nervous system prioritizes and protects the heart and lungs. The brain has an incentive to ensure that these two organs are well protected, since glucose travels in the blood. Thus the brain will send a signal to the rest of the body saying, in effect, "On average, we are only receiving four glasses of water, or one-third of the daily water required, so I am only going to give you all one-third of the daily water you should be consuming." Your brain will then orchestrate a precise plan to help the rest of the body cope while it draws water from other parts of the body for its own survival. Its only recourse is to steal water from less-important organs and systems.

Chapter Eight

The Order of Things

The only difference between stumbling blocks and stepping stones is the way in which we use them.
—Adriana Doyle

Consider forests and farmland. In times of abundance, forests and farms are commonly adjacent to rivers or lakes and bathed in the warmth of the sun. Forests are quite noisy places, replete with calls of various species of birds and the chattering of squirrels. The wind rustles the tree leaves. The smell of earth, flowers, and moisture fills the air. The plants are vibrant with color. Farm fields teem with plant life. Birds, insects, and the beauty of nature as a whole make their presence known everywhere. Nature always performs in perfect harmony.

Imagine now a changing scene—one in which water has become less readily available. As forests and farmlands are deprived of water, to the point of complete absence, plant life moves into survival mode. Plants and animals are increasingly unable to protect themselves from the sun's relentless heat. Every living thing in the forest reacts to the scarcity of precious water. Various creatures will approach drought differently, but all will most certainly react.

Grasses turn yellow and sometimes fall to the ground. Gardens and shrubs deteriorate. Trees and their branches appear limp and distressed. In addition, there is an increased population of insects as stressed plants make easy targets for them. This transformation takes considerable energy. Similarly, it takes a lot of energy for us to keep our core at a safe temperature in extremely hot conditions. The body defends itself by distributing stress so that no one part of the body has to bear the brunt.

The Skin

This is the first organ the brain will head for in its pursuit of water. Skin is the body's largest organ. It is one of four toxin-dispensing organs. The lungs, bladder, and colon are the other three. The skin is also responsible for up to 15 percent of your oxygen intake. Your lungs supply the other 85 percent. Signs of dehydrated skin are excessive peeling, irritation, and a greyish hue.

Temperature is the most important function of homeostasis. Homeostasis can be considered the human body's thermostat. It regulates body temperature, digestion, defecation, blood circulation, respiration, nervous system and immune system, among other duties, mostly involving maintenance and repair.

In wintertime people will say, "My skin is so dry. It must be from the electric heat in my house." However, for the most part, it is the cold outside that is responsible. You lose water, which is the form of fuel required to efficiently keep the body core heated. We drink more coffee, tea, and hot chocolate in this time of the year as well. Rarely do we reach for cold water. Our bodies are being robbed of precious energy. Extreme inside-outside temperature contrasts cause the body to burn more water to maintain its core temperature of about 98.6 degrees Fahrenheit. These are all areas to which we pay relatively little conscious attention.

Lymphatic System

Once the brain has sufficiently redirected water from the skin for its own requirements, it turns its attention to the lymphatic system. Lymphatic fluid is 95 percent water; it touches and surrounds every cell in your body. It rinses every cell on a continuous basis to help clean and restore balance. The lymphatic system also transports contaminated lymph fluid for disposal. In fact, the lymphatic system keeps things clean so that the cells are ready for the next biological function. Using lymphatic canals, lymphatic fluid normally picks up toxins and processes them through the lymph nodes or storage glands. Once this fluid has been processed, the lymphatic glands cleanse themselves by unloading waste into the lungs. The lungs in turn process it further and then dump the waste into the colon. I think we all know where it travels from there.

For this reason, dehydration can play a role in the development of breast or lymph cancers.

For the most part, cancer cells thrive in an environment that contains very little or no oxygen—an anaerobic (without oxygen) environment. If you allow the canals of the lymphatic system to dry up or to move fluid sluggishly, your lymph glands will be full of toxins and stagnant fluid. This can leave glands around the breast dried and clogged, an anaerobic state in which cancer may develop. The body is unable to flush the lymph glands properly, thereby preventing the lymphatic system from doing what it's designed to do.

Water is two parts hydrogen and one part oxygen (H_2O). Water not only helps produce ATPs, it also channels oxygen at the cellular level, which could decrease the chances of cancer cell development.

This does not mean that drinking water alone will prevent or cure cancer. But if you don't drink water, over time the development of anaerobic conditions creates a perfect breeding ground for cancer cells. Someone who works out regularly, especially aerobically, forces the toxins into the muscle tissue. The lymphatic system then cleans them up and clears them out.

If you open the floodgates using water for thorough hydration, you will stimulate a constant healing process.

> *The flowers of the monsoon rains in the desert will*
> *be yours to have and experience.*

The Muscles

Once it has relieved the skin and lymphatic fluid of all the water available, the brain then starts to look for another large system to contribute. Muscles are next on the list. All muscles will be affected by dehydration. Water corrections will continue throughout the various systems of the body. Energy/water loss will cause muscle atrophy. If your body is starved of energy, muscles will begin to break down in order to maintain safe energy levels. Visualize a dehydrated orange or plum. Chronic dehydration will have a similar effect on your muscles.

Excessive lactic acid deposits on the muscles will also occur. Lactic acid is naturally expelled from muscles when they are worked or stressed. A lengthy, spontaneous game of golf or a long, unplanned bike ride, for example, can cause nagging low back or neck pain. The lymphatic system usually plays a major role in cleaning that up, but it is running at about 40 percent of its capacity and attending to more crucial issues

Try to recall a time this happened to you: You played beach volleyball or went golfing or biking for the first time in a long time, and the next day you experienced a severe reality check. Your weak, dried-up muscles were damaged. You had become accustomed to very little cardiovascular activity. As a result, your lymphatic system was not up to the task of cleaning your cells, and your immune system was incapable of rebuilding the damaged muscle tissue. When we are young, we have the resiliency needed to overcome most conditions and to operate generally free of pain. As we age, it becomes painfully obvious that our bodies can no longer tolerate this type of physical abuse. For most of us, there will come a time when we realize we're not invincible. Typically, this realization hits us between thirty-two and thirty-seven. Certainly we're well aware of it when we reach forty years of age or older. I've been told it doesn't get any better.

When you have conditioned your muscles to look, perform, and rest in a systemic, dehydrated state, they will be seriously challenged to deliver power and flex with the ease you might have expected. The length of the muscle is drastically shortened. It also suffers the effects of poor lubrication. This will deny the muscle the power required to complete the chore at hand. Exertion will cause extensive microscopic muscle and ligament tearing, creating scar tissue. This can happen even in an ideal environment, but to a much lesser extent.

In my years of clinical practice, I have found that stress headaches are very common and that the main trigger for these headaches is dried-up suboccipital muscles. These muscles are located at the base of the posterior (back) part of the skull. They are attached to the first two vertebrae (C1 and C2). The main function of these small but important muscles is to allow small movements of the head or skull. They are used extensively when you sit in front of a computer, when you drive, and when you read—whenever you engage in activities that require you to perform relatively small movements of your head, especially for extended periods of time. They also assist in stabilizing the skull to your upper spine when jogging, for instance. When your body is in a condition in which it is constantly searching for energy, the brain will disengage the large muscles of the neck and thrust all the responsibility onto the suboccipitals.

We must accept responsibility for how we feel. Medical science blames heredity for certain illnesses. But in reality, issues you may be dealing with that are shared by parents or siblings are usually directly due to similarities in nutrition and lifestyle.

Take arthritis, for example. I once owned a lawn-care company, and when I had to explain moss control to a client, I would approach him or her with this logic: I would tell the homeowner that there is no way to eradicate the millions of moss spores in the soil on his or her property. An attempt to do so would prove very expensive and would involve a total soil removal and replacement job. Moss can survive and thrive only in an acidic environment. That is to say, only when the soil's pH level is extraordinarily high. One thing grasses thrive on its soil with a 7.2 to 7.6 pH level, much the same as is in our bodies. In highly acidic soil conditions, the grass plant will weaken, become thin, and perish. This is not for want of minerals in the soil but for an inability to process those minerals due to weaknesses brought on by the acidic soil. On the other hand, moss loves acidic soil; it can't survive without it. In acidic conditions, moss will aggressively consume the growing area and eventually eradicate or choke out the grass. The solution then is to reduce the soil's pH level to an acceptable level for grasses by using generous amounts of an alkaline, such as lime. This allows the grasses to thrive and, in turn, discourages moss growth.

At this point you may well ask, "What does moss have to do with arthritis?" Arthritis thrives in an acidic environment; by maintaining a proper pH level, most arthritic diseases can be discouraged.[7]

---

7   Howell, Edward. *Enzyme Nutrition*. New York: Avery, 1985.

Think of someone you know who has arthritis, fibromyalgia, chronic fatigue syndrome, or digestion problems. I suspect that they will rarely reach for water as a thirst quencher! Chances are their beverage of choice would be coffee, soda or soft drinks, alcohol, or large quantities of apple juice, orange juice, and lemonade. Others might have little to no fluid intake at all. Their food preferences would generally have a tomato base with black pepper and red meat, all of which are acidic. They likely eat very little alkaline food and, of course, little or no clean water. Such a style of food consumption sets up the perfect acidic level to promote and nurture arthritis.

I love the subtitle of Dr. F. Batmanghelidj's 1992 book on human hydration, *Your Body's Many Cries for Water: You're Not Sick, You're Just Thirsty; Don't Treat Thirst with Medication*. The truth behind it is quite overwhelming. It is simplicity at its finest. In this book, the author, a physician, explores the physiological and pathological effects of dehydration in detail. Water is a great alkaline. It assists the body in reducing acidity. Calcium, magnesium, and calcium-rich foods (all dark green vegetables) are natural alkaline foods that help reduce pH levels.

We now realize that most of our muscle pain is because we have allowed our muscular system to dry up. This sets up an acidic environment, which aids in arthritic development. In addition, a lack of cardiovascular activity will stress and burden the lymphatic system. This is because muscular activity is the pump that drives the fluid through the lymphatic system.

We also expect the muscles to perform at peak efficiency, even under these high levels of acidity and dehydration, conditions that impair the body's ability to transfer and synthesize vitamins, minerals, and nutrients. Top performance efficiency is simply not going to happen in these circumstances. Not if you are human and physiologically normal. Instead, you are going to micro-tear extensive amounts of muscle tissue. This will cause scar tissue to form, which will eventually create a repetitive stress injury. It does not stop there, and it won't until you hydrate.

Chapter Nine

What Your Organs Have to Say

> *Is there anything worse than being blind? Yes, a man with sight and no vision.*
> —Helen Keller

Now we will discover how the internal organs are affected by dehydration. First, let's consider the colon as the brain scrambles to find more precious water during what it senses is a drought. It will instruct the kidneys to extract water from the colon, purify it, and send it up. Such a physiological response will translate into periods of constipation or diarrhea. Over 80 percent of cancers either begin in the colon or exist because of a malfunctioning colon. I have a special place in my heart for this organ.

For years I have lectured on the topic of healthy pooping habits. Don't worry; I won't give you the long version. These are just the facts. If you are eating two to three major meals a day, you should be experiencing at least two daily bowel movements of the 4F calibre. A 4F bowel movement is one that is fawn, fluffy, foot long, and a floater. Rabbit-type movements indicate that you are constipated. For those of you so afflicted, there are usually two solutions: take in more fibre and drink adequate amounts of water.

In a state of dehydration, the kidneys will be instructed to withdraw water from the colon. This will cause the bulk matter in your intestinal tract to dry up, slowing down ingestion, digestion, and toxin disposal. This, in turn, causes almost all the oxygen to be squeezed from the colon to be replaced with layers of toxic anaerobic fecal matter. That's why the colon is responsible for playing a role in more than 80 percent of all cancers. Like moss, cancer can readily flourish in an environment that is devoid or nearly devoid of oxygen. Remember your childrearing days? When you fed your infant many carrots, orange would appear on the skin and in the diaper within four to twelve hours. We were born with our colons functioning normally like that. Your colon is your body's best defence against toxic substances.

The common reaction to food poisoning is abdominal cramps. This may be followed by vomiting and/or diarrhea because the point of entry of the poison was through the stomach. Even extremely constipated individuals will experience a wild session of diarrhea under these conditions. As a general rule, almost all organs and systems in the body discharge toxins directly or indirectly to the colon.

When diarrhea or constipation take hold, the colon is signalling the liver, kidneys, spleen, lungs, and lymphatic system that the dumping doors are now closed or severely restricted. The colon is having a problem dealing with its most basic function, that of processing food and dispensing toxins. This will then force the other organs to find alternative methods of dealing with these toxins.

The liver will dump its toxins into the blood because the colon is shutting the door. The liver is filling up with stored toxins. Just like a glass of water will overflow if you continue to fill it up, the toxins in the liver will overflow into the blood. The skin then has to try to dispense the toxins, doing so through pimples and boils. The skin will appear grey in color.

The bladder will be forced to deal with a high toxicity level as other organs and systems divert their waste to it. This in turn increases the stress on the bladder.

The workload of the kidneys will increase because of the rise of toxicity levels in the blood. These organs will be stressed just by reallocating the water.

Since the lungs are also one of the toxin-dispensing organs, mucus production will increase. When levels of toxicity rise, the body relies more on its ability to entrap the toxins by surrounding them with mucus to demobilize them until they are transported out of the body. This physiological reaction adds to the totality of stress. More water is required to produce and move the increased amount of mucus.

Certain digestive problems may develop that could result in heartburn and stomach issues: ulcers, colitis (inflammation of the bowels), diverticulitis, Crohn's disease, headaches, anxiety, and so on. Your dehydration habit is now interfering with both your body's ability to utilize the food you're consuming and its ability to expel fecal matter.

When your brain is finished with the colon, the kidneys will be the next organs targeted. The liver and the spleen will follow. The kidneys are the first organs seriously affected during the primary stages of dehydration. One of the main functions of the kidneys is to police the distribution of water, sending it to the most vital regions first.

You may be reading this and thinking, *I must be a camel because I abuse myself and I still feel great*, but do not delude yourself. You won't know how good you can feel until you have hydrated for at least three months. The brain will distribute the extra water in the same order of priority it used when robbing the body of water. Most people will feel the most noticeable difference when the muscles are hydrated. The last organ to be replenished will be the skin, which is the first organ from which water was withdrawn when dehydration began. You will experience some benefits before three months, but that period is about what is needed for the body to regain its balance.

Chapter Ten

Weight Loss and Water

*Persistent people begin their success
where others end in failure.*
—*Edward Eggleston*

Throughout my clinical years, I have seen the debilitating effects of dehydration in the everyday lives of my patients. Dehydration affects the body's ability to gain weight or lose it, but not in the way you may think.

Dehydration and hydration are not simply weight gain, weight loss. There are many contributing factors to successfully losing weight, but by far, the main factor is water. All successful weight-loss companies advise their clients to drink ten to twelve glasses of water daily. Their clients are coming to them in a dehydrated state, which means their bodies are in a state of starvation. As mentioned before, in starvation mode you lose weight in the beginning. Then the body goes into defensive mode due to lack of water and there is weight gain because the body will make you tired, slow you down so you don't waste energy, take whatever solid foods you are consuming, and store it as fat. Fat storage is a defensive mechanism against starvation.

I have tried weight-loss programs with patients in a variety of ways:

- the group method, which usually ended up in debates
- the buddy system, where two friends support each other—the most successful
- the one-on-one approach

I found that one thing is universal. It isn't lack of information; the patients know what they have to do. The fact is that they're addicted to the food that makes them fat. As an individual wanting to lose weight, sooner or later, you will have to deal with this demon. In most cases, just subtle changes are needed to get to your target weight. My approach is to support the systems in the body that are required to accomplish the task.

Water First, of Course!

It's not as simple as just drinking three litres of distilled or clustered water every day; you need to eliminate the diuretics. As mentioned before, you can substitute the diuretics with herbal teas that support the system. Some of the herbal teas that help support the body during weight loss are ginger, dandelion, liquorice or anise, peppermint, and ginseng. And all of these count toward your three litres. I recommend drinking a variety of them. I would choose dandelion as one of them (a powerful liver detoxifier). You need to condition yourself to drink every half hour to an hour if you can. The body needs an energy supply on a continual basis. It is natural for a properly hydrated body to urinate four to six times a day. Let's not forget the other functions water plays in ridding the body of fat. The body stores toxins as well as the sugars in the fat, so when you start losing weight quickly, you will also increase your toxicity level. A hydrated body can handle that. Drink water regularly during the day. It will turn off your hunger signals, assisting the kidney and liver and colon to break down fat and dispense toxins.

Another organ that needs support during weight loss is the heart. I have my clients take high-end EFAs, essential fatty acids, also known as Omega-3, Omega-6, or Omega-9. They are called "essential" because it is essential that we add them to our diet. Our bodies can't manufacture them; therefore we have to consume them. Many EFAs are found naturally in raw meats and some raw vegetables. You can find reputable natural products on the market. Caution: you don't go to a health food store to fill your prescription drugs, so don't go to a drug store for anything truly natural.

Why EFAs?

Take a moment and glance at the size of the palm of your hand. That represents the size of your heart. This little pump pushes outward into a network of approximately six miles of arteries and arterials, from vessels that are large to hair-like in diameter. The heart distributes food to all the cells in the body, helps control heat, and helps with muscle movement.

EFAs lubricate all the vessels in your body. Some of them are the arterials and the veins; this is how the blood returns to the heart to be oxygenated, through the venous system. This is also referred to as blue blood, which means blood without oxygen. Arterials, veins, colon walls, and chambers in your heart and lungs are all affected. If your diet lacks natural EFAs for a long period of time, as in the typical cooked North American diet, the reverse happens. All the vessels in your body become rigid, and those little hair-like arteries dry up first. This in turn causes back pressure to the heart. Inevitably, this will result in high blood pressure and/or high cholesterol problems. Most people on high blood pressure medications arrive at that point mainly due to the lack of EFAs in their diet. Therefore, introducing good-quality natural EFAs can radically cut back on or eliminate the drugs completely.

My father passed away of a heart attack at forty-nine years old, and four of his brothers died around the same age. Looking back, I know now we didn't have a lot of natural EFAs in our diet; even though we ate lots of fresh fish, it wasn't raw. (Note: If you apply heat to those EFA foods, you destroy the healthful properties of the EFAs.)

I have had many clients reverse their high blood pressure by the introduction of EFAs. Another misconception is that fat makes fat. Raw sugar makes fat. There are good fats and there are bad fats. People tend to ignore the good fats along with the bad fats. EFAs are one of the good fats. EFAs can readily be found in these foods: raw fish and shellfish, flaxseed (linseed), hemp oil, soya-bean oil, canola (rapeseed) oil, chia seeds, pumpkin seeds, sunflower seeds, leafy vegetables, and walnuts. The liver requires ample good fats or EFAs to assist it in the burning of bad fat. Taking in a bad fat, such as animal fat, doesn't make you fatter; it just raises your cholesterol levels, giving you fatty blood. Bad fat, another toxin, will be stored in your body's food bank.

In Dr. Timothy Brantley's book *The Cure*, he outlines this extensively. His book shows the enormous role EFAs play in the fat-burning process. To be nice to that little muscle called a heart, just add the needed supplements; after all, it asks little of you. The body will do the rest.

The Colon

This, of course, is one of the most important organs involved in weight loss. I have previously explained the colon's function at length already. The whole process of burning fat and disposing of the body's toxins is severely compromised when your colon is sluggish. Again, supplement fibre intake is indicated. I had my patients in the weight-loss program take fibre before every meal. Taking fibre before each meal accomplishes two things: It arrives before the meal to scrape and clean the small intestinal tract where the beginning of digestion happens, and it makes room in the large intestinal tract for the coming meal, making you feel half full; as a result, you do not overeat. Since the fibre is assisting in proper digestion, you will extract more nutrients from what you eat.

Bee Pollen

Bee pollen is yet another marvel from nature. Albert Einstein once said, "If all the bees died, mankind would die of starvation within seven years." Those bees work without holidays, every day moving pollen from one plant to the other, cross-pollinating to ensure the strength and fertility of all the plants they come in contact with. The pollen they gather is all they eat. Bee pollen has all the vitamins, minerals, trace minerals, digestive enzymes, and amino acids (building proteins of the body) and is the only natural form of folic acid; it is nature's perfect food. Because pollen derives from the same DNA pool as we do, the body recognizes it and is able to utilize all of it. I recommend bee pollen along with fibre and water to help children with attention deficit disorder. I recommend my patients take bee pollen two to three times a day while dieting. This helps to ensure you are not malnourishing yourself by depriving yourself of all important trace minerals and vitamins; the digestive enzyme will help your body digest cooked foods, which are void of digestive enzymes. Bee pollen is an amazing fuel for your immune system; everything in it is the fuel needed to assist in repairing things properly.

Exercise

If you simply employed the three previously described changes without doing anything else, you would start losing weight. If you introduce simple cardiovascular exercise to your program, then the process of burning fat will be more pleasurable. Instead of just burning the fat, whose main by-product is toxin, you will also burn the sugar the body was searching for. So a cardiovascular activity for as little as fifteen minutes a day will help the body flush the excess toxins and burn more sugar. Types of cardio activities include swimming, biking, walking, and rowing.

I personally avoid jogging. Jogging is the only activity that eventually you will have to stop doing because of repetitive stress injury on every joint in your body. You don't have to quit swimming, walking, or any other activity. Long-term jogging will eventually lead to foot, ankle, knee, hip, low back, and/or neck pain, which leads to constant headache.

Avoid manmade carbohydrates or gluten—better described as glue once it enters the body. A lot of great literature has been written on that; the Atkins diet and *The Cure* will enlighten you further.

Implement those easy things and fat will fly off you! The fact is that 80 percent of you will not apply all the steps, just the ones you like. The others are too painful.

Let's review

- ***EFAs*** (essential fatty acids Omega-3, Omega-6, and Omega-9). Good for the heart, these increase distribution of food and oxygen to every cell in the body and help the colon dispense of toxins.
- ***Exercise.*** Cardiovascular exercise increases lymphatic, cardiovascular, and venous flow, which cleans the cells in the body.
- ***Colon***. Fibre helps the colon to function at peak performance to enable it to move unwanted material out of your body.

- *Bee pollen.* This feeds the immune system, which in turn increases your energy and your focus, and it helps cut the cravings between meals.
- ***Avoid carbohydrates, eliminate gluten, and increase your clean protein.*** Stop putting glue and hordes of sugar in the form of carbs and sweets into your body. Seek some protein foods and raw foods you like.
- *Water.* Drink three litres (twelve eight-ounce glasses) of distilled or clustered water per day. This makes it possible for all the above factors to help your organs reach their highest level of efficiency.

Follow those instructions, and you will safely lose the weight you want to lose. As with any new diet and exercise regimen, you should consult your physician before beginning. Neither the author nor the publisher shall be responsible or liable for any loss or damage arising as a consequence of your use or application of any information or suggestions contained in this book.

Chapter Eleven

You and Your Unborn Child—Pregnancy

*Life's tragedy is that we get old too
soon and wise too late.*
—*Benjamin Franklin*

You don't have to have a lot of common sense to understand that water deprivation during pregnancy is harmful at many levels, both to the fetus and to the mother.

A few things that are unique to pregnancy happen to the female body. The first thing that happens once the egg is fertilized is that hormones go crazy within the body, especially estrogen, because this hormone reproduces cells, even fat cells. After that, the female body's only purpose is to produce a healthy fetus, even to the point of sacrificing her own life for the survival of her child. Now imagine you were very chronically dehydrated and still able to get pregnant. The next nine months would not be pretty if you were trying to do it without precious water. The body has no problem sacrificing itself to get water to the fetus.

Second, the human body does not have the capacity to store water. Only during pregnancy will the body store water in the space around the outside of cells (in the interstitial fluid). The body will deposit precious water to the feet, then the hands. I am sure you have heard of women who get swollen feet and hands during pregnancy. The body is storing water there for the fetus because your body knows your drinking habits. The body stores it in the extremities because it does not go there for water when it can get it elsewhere by expending less energy.

The initial response might be, "Oh my God, I am retaining water!" The pharmaceutical solution is to stop drinking water and take chemicals to deplete this storage bank. This only happens to certain women because they are trying to create the miracle of life without sufficient clean water. The pregnant body goes through a major growth process as well. Most women feel better when they are pregnant, assuming their diets contain adequate nutrients and they are relatively hydrated. The female body will move into a detoxifying mode, meaning the immune system is bumped up a few notches. The body is cleaning house for the new arrival. By now you realize that detoxifying anything in the body requires lots of free water. The feeling of wellness comes from a woman's body operating at a more efficient level. Normally pregnant women won't catch a cold or the flu due to the immune system operating in high gear.

As I explained earlier, the blood-brain barrier is a very efficient blood-filtering device. Its main purpose is to ensure the blood is not contaminated when the brain feeds on it. The fetus also has Defence systems. The placenta and surrounding walls are very efficient filters. The female body filters everything that has to do with the fetus. It is imperative that there is enough water flowing through the body to ensure that toxins don't interfere with the process of development.

The dehydrated pregnant female puts additional stress on the fetus. The higher levels of toxins in the woman's body interfere with basic functions of the body during this important time. As explained prior, the small intestinal tract extracts minerals and vitamins from the food through receptors. Toxins block these receptors and because of that, there's a lot of deprivation to the body. This deprivation is transferred to the fetus's growth and development.

Going through the pregnancy hydrated, the mother would not have to experience symptoms of constant fatigue. Dehydrated women will feel drained physically and emotionally, particularly in the third trimester. They could avoid 90 percent of their muscle pain if they supplied their body with enough water. At this time, the supple abdominal cavity has more elasticity to it; it is able to stretch with less resistance and less muscle tearing and, consequently, less pain. But if the abdominal cavity is dried up, it resists any movement, and the result is a lot more muscle tearing and a lot more pain.

Practicing yoga or Pilates during pregnancy can aid in reducing muscle pain. Yoga is the art of stretching. Pilates is another discipline worth practicing, even before you get pregnant, since it is a core-building art, and the exercise strengthens your abdominal muscles. Your abdominal muscles will be undergoing the most change. The stronger your muscles are as your abdominal wall is forced to grow, the less low back pain you will suffer. Weak abs won't be able to carry the extra weight, so the back will need to take the majority of the workload.

Morning sickness during pregnancy can usually be eradicated with proper hydration, along with heartburn, bloating, edema headaches, indigestion, insomnia, depression, and stress, to name a few side effects.

If you're pregnant right now, please consider drinking clean water daily for your unborn child. Hydration plays a part in your development months before you take your first breath. Remember, "Dehydration is the disease."

Chapter Twelve

Reversing ADHD without Drugs

Yesterday is history. Tomorrow is a mystery. Today is a gift. That is why we call it the present.
—*Eleanor Roosevelt*

We have seen influences water has had on us during pregnancy. The next stage of life is growing up. What is important when your body is in that first major growth spurt? From the womb to roughly fourteen or fifteen years of age, the body increases the amount of growth hormone it produces until the growth spurt is over. There will still be further development into adulthood, but the rate of growth will be less noticeable. This age is also the time the child can be taken off pharmaceutical drugs for treating attention deficit hyperactivity disorder (ADHD). Children generally grow out of it by their early teens.

It is hard to conceive that a condition like ADHD, which is prevalent in North America, can be corrected safely and naturally with diet and water.

ADHD is classified as a disorder. In my opinion, the cause of the disorder is malnourishment and it is totally reversible, naturally. In my years of practice, I have treated many children with ADHD. It is a three-session program. The first visit lasts from sixty to ninety minutes. I counsel only the parents, and the child is not present. During this session I educate the parents on the disorder. The disorder stems from insufficient amounts of living nutrients in the child's diet. Not drinking enough clean water is a huge part of it. In most all cases that I have treated, the children's fluid intake consists of sugared juicy cups, fruit juices, soda or soft drinks, and milk; if they took in water it was tap water.

Surely by now you have learnt enough to know this group is missing out on the most precious food, water, whilst the body is experiencing its biggest growth spurt ever. This is why it is equally important to supply the proper nutrients in the solid food intake. So besides educating the parents on hydration, we review everything they bring in the house for the family to consume.

The most expensive aisle in the grocery store is the junk-food aisle. They are forbidden to go down that aisle for the future health of their child. This in turn frees up a lot of money to spend on real food. I suggest they go gluten-free immediately and remain that way for at least three months. Remember, it takes 120 days for your body to replace all cells. After one month, they must eliminate all processed foods. At the beginning of the third month, we cut back on cooked food and eat more raw foods. Again, Dr. Brantley explains this in detail and offers countless recipes to waken your mind to new meals, meals that are alive.

If the parents had a hard time with those rules, chances are the drinking rules will likely not be welcomed with a smile. Most of what I tell them in that first ninety minutes is received with a look of shock. These are the drinking rules: they are advised to offer only water or herbal teas to their child. Make the tea and add it to the water for flavour and variety. No more fluids that carry sugar, caffeine, or artificial ingredients. Their child is trying to build a body to be proud of, not made of artificial material.

The parents are instructed to lead by example. Purchasing a cooler or crock to hold eighteen or nine -liter bottles, or 5 gallon or two and a half gallon containers of distilled water helps the cause. Observing Mom or Dad drinking from the water dispenser will have a positive reaction; the child's instinct to mimic Mom and Dad will be triggered, and he or she will start drinking water too!

Adjusting diet for the whole family allows the parents a wake-up call. After all, they eat the same foods. Parents must realize and then take responsibility for their child's disorder. This is not meant in any way as a judgment, and taking it personally would be counterproductive for everyone involved.

Once we are clear and in agreement of the new rules, together we set up the second meeting for about ten days later, at which time I will meet the child. The parents go home and tell their child the new rules. I instruct the parents that at no time should they reinforce or even mention ADHD, particularly in a negative manner. The teachings should be channelled to a productive, positive approach.

It should be noted that these children are usually very intelligent, but the condition impairs their ability to sort out incoming information. The parents will go home and tell their child that they went to a counsellor who helps children become smarter in school and that these rules are the consellor's rules and that in ten days he or she will be meeting this counsellor. The parents should also assure their child that he or she is not doing it alone. The whole family will be doing it. They should reinforce the fact that the child will be doing better in school, because that is what the child is struggling with the most. As a result of this he or she experiences self-doubt, frustration, anger, and all the behaviours that led to him or her being put on the drugs. The child doesn't want to be singled out that way; he or she wants to be one of the normal kids.

During the ten days, if all is implemented, there will be a noticeable difference in the child by the time they bring him or her to me. That is the main reason I wait till the water and food has been in the child's diet and the fake food is no longer being ingested. The child will be more receptive to what I will be teaching him or her.

Normal levels of cognitive thinking are often distorted in ADHD. Because the child's diet is nearly void of proper nutrients, it is very hard for him or her to concentrate. Without proper electrolytes in your diet, the synapses in the brain are unable to function normally. This activity is how our brain cells communicate with each other. This is why these poor children are struggling in school, where cognitive thinking and skills are developed and challenged. It has been my experience that when the diet is changed, the child's behaviour changes. If the child is already on ADHD medication, after three to four weeks, there is no need for the drug.

In my program, I also check the eyes to see if they have parasites; if so, I treat them. An infestation of parasites can cause the child to have sugar cravings, and the parasites are consuming much of the nutrients in his or her diet. The other recommendation was to put their child on bee pollen to ensure that he or she has adequate minerals and vitamins to supplement his or her diet. I tell the child that the bee pollen is brain food and that doing homework, writing, and reading will be easier. I also introduce a quality EFA.

Before I meet the child, he or she has had ten days of healing. After my visit with the child, we schedule another appointment fourteen days later. This allows almost a month of change, and the child is always noticeably more focused. We take this opportunity to have a question-and-answer period. I suggest a list of reading materials to support my claims. The more educated they are, the more educated their child will be. The door is always open for the future, but three meetings usually suffice.

I attended an herbal conference in the United States, and one of the seminars dealt with this topic. The speaker was formerly a heart surgeon. He left that profession to work for a pharmaceutical company. During his tenure there, he was on a team with other scientists, and together they developed a drug for treating ADHD. After a few years, he left the company and went to the other side; he worked for a major herbal company.

In his lecture, he explained that diet was the main cause of the disorder. When he was done speaking, he accepted questions from the audience. One question posed to him was, "What exactly is the drug and how does it work?"

His response was shocking. He said, "There is no difference between this drug and street speed. The drug keeps the child in a stoned, subdued state, and their logic is the child is calm now." He added that after two to five years on the drug, the child will develop an addictive-type personality, and other stimulants will easily hook him or her. The doctor went on to say he wasn't proud of his time at that company and was on a mission to educate people on the natural way to heal the body. His recommendation was to introduce a supplement fibre to assist the colon in ridding itself of gluten-type foods and other toxins and using bee pollen to address the deficiency. It is always about imbalance; it is all about not being natural.

Chapter Thirteen
Reversing Fibromyalgia Naturally

> *The first step in the acquisition of wisdom is silence,*
> *the second listening,*
> *the third memory,*
> *the fourth practice,*
> *the fifth teaching others.*
> —Solomon Ibn Gabriol

Fibromyalgia is a medical condition that in my opinion should be the flagship for water awareness.

What Is Fibromyalgia?

Fibro means "muscle" and *algia* means "pain," so loosely translated, the word means muscle pain. In my opinion, it should be called toxicalgia, because it is pain derived from the presence of high levels of toxins over an extended period of time and not from the muscles.

The muscles are just talking to you the loudest, and in most cases, they become so weak that they can't carry their skeleton. If learning about this condition does not motivate you to drink more clean water, nothing will.

Fibromyalgia is a relatively modern condition. It has been brought on with the baby boomers, and now the baby boomers' kids are getting fibromyalgia at alarming rates and at an earlier age. If you have been diagnosed with fibromyalgia, then

- there is an 80 percent chance you are female;
- you are most likely a type-A personality (most type-As will deny they are); you are juggling twice as much as any of us would want and never do anything for yourself.
- you would have had to deny yourself adequate amounts of clean water for about fifteen to twenty years;

There are existing conditions that might throw a person into the condition quicker. Fibromyalgia can be brought on by severe whiplash or a traumatically stressful period. After two years of the body trying to recover from the destruction caused by the whiplash or high levels of stress, the body will set aside some of its efforts to battle toxins, and all the available energy is sent to deal with the bodily assault at hand. During this time, the body's toxicity levels reach another high, and the body can hardly feed itself, let alone do a major cleanup. When Western medicine looks at the symptoms, you are diagnosed with fibromyalgia.

The individuals whose fibromyalgia was triggered by an accident or chronic stress were most likely predisposed to the disease prior to the incident. These people have dried-up muscle tissues that will tear with little resistance. The triggering incident just amplifies the amount of damage done to all muscle tissue within the body.

Case 101—Danielle

You need to know what a person goes through when dealing with fibromyalgia, so at this point I would like to introduce Danielle, a website guru. I met Danielle through a mutual friend. I needed a website, and she had fibromyalgia; we were a match. When I first met Danielle, she was thirty-four, and she came in draped over a cane. Without the cane, her knees were not strong enough to hold up her hundred-pound, weakened frame. At this point on her journey, she had already cleaned up her diet; she was juicing and eating raw—always a healthy approach to detoxifying the body. She was still losing weight; she couldn't afford to lose more! Danielle had raised her body's toxicity level so high that most of her cellular receptors were blocked, hence negating the cell's ability to feed itself. To counter this, the body breaks down protein or muscle tissue in order to feed itself.

On Danielle's first visit, she struggled toward the couch. I asked her a question to which I already knew the answer: "Would you like a glass of water?"

She politely replied, "No, thank you."

"Wrong answer," I replied, and I proceeded to pour her a large goblet of distilled water with a wedge of lemon. Remember this was our first visit, our first words… she said she felt like she had just been told! As I handed the water to her, I said, "This is why you are sick." Of course, I did not leave it at that. I told her if she did what I asked of her, she would lose the cane in only two weeks. In her case it took three weeks.

I spoke with Danielle on that first visit for one and half hours. In that time, she drank two of those goblets of water—about a liter and a half, or six glasses of water. She went to the washroom five times to urinate over that ninety-minute period. Her body was not capable of absorbing water at first because it had been conditioned to operate without clean water. That is usually why individuals with fibromyalgia will give up on trying hydration; it is too painful to move, and it is just not worth it for them.

Danielle agreed to enter my four-month plan to reverse fibromyalgia. That is how our friendship started, and we continue to work together today.

To manifest fibromyalgia, these individuals turn off the clean water. Therefore, to reverse it, the first step is to turn *on* the clean water. The people fighting this disease have to work hard at avoiding water and consuming diuretics for energy. As in Danielle's case, the development of the disease was slow and insidious. Inevitably the patient will be focusing on the symptoms or simply ignoring them, which is a type-A personality trait. The focus is put on symptomatic responses and not on the real cause. The cause is often silent; stress and dehydration are the top two examples, but the result is total systems shutdown.

My approach to defeating or reversing fibromyalgia is an arsenal of action against toxicity in the body. As mentioned earlier, there are only four toxin-dispensing organs in the body. To follow are explanations of how each of these dispensing organs function naturally and physiologically in your body. I will also point out how the body would react if the dispensing organ was malfunctioning as it does with fibromyalgia.

The colon assimilates many types of toxins, and other organs dump their unwanted waste into the colon for processing and dispensing. When this organ is exposed to chronic dehydration, it dries up, unable to efficiently assimilate food. It slows down and stops, naturally pumping the fecal matter forward. In the pecking order of when the body goes hunting for water, this organ is number four on the list. When this organ is weak, parasitic infestation is quite common, especially in this day and age, with global food, travel, and trade.

Danielle agreed to a three-month parasitic cleanse using herbs, fibre, and of course, water. She passed at least six different kinds of parasites. That is why, even with juicing, the easiest way for the body to assimilate food the parasites and toxins were negating proper ingestion. In Danielle's body, the toxins were intercepting the cells' gates and receptors. The cells weren't eating sufficiently, and the parasites thrived in the abundance of available food. It's no wonder she was weak—her body was starving!

The second dispensing organ is the skin, the largest organ in your body. The skin is number one on the hit list when there is a shortage of available water. The skin is a breathing, ever-evolving, outer layer to our physical form. Along with fascia, the skin assists in holding us together. It replaces itself roughly every four weeks. Chronic dehydration negates the body's capacity to utilize the proficient healing ability the skin has to offer. The skin takes in roughly 15 percent of your oxygen, and in the process, it filters toxins from the air and feeds the body.

If you have been on a detox and found that you broke out in acne, your skin was doing its job. If the liver is being detoxed, it will dump some of its toxins back into the bloodstream, since the liver is the recycling center of the body. Everything will be processed there eventually, and this organ works with blood. Once the toxins are blood bound, the skin is an easy way for the body to cleanse the blood. In all my fibromyalgia cases, there was an obvious reaction from this dispensing organ. As mentioned these individuals are often grey in skin color, indicating poor circulation. These individuals' livers are extremely overworked, and their blood is very dirty. This is why they always break out in boils, usually on their backs. With the presence of water, the skin is able to help clean the blood. After roughly five weeks of hydration, the skin rejuvenates itself.

The third dispensing organ is the lungs, assisted by the lymphatic system. Remember, the lymphatic system is the second place the body searches for water. It literally rinses around every cell in our bodies. Once the lymph nodes have processed the lymph fluid (which is 95 percent water, carrying toxins), they send it toward the lungs through a network of canals. The lungs will filter it further, using evaporation to dissipate the bulk of it, and the colon will inherit the rest, or we will cough up mucus. Serious imbalance occurs when these lymphatic rivers begin to slowly dry up, and mucus membranes are unable to complete their work.

A deficient lymphatic system in turn passes on a lot of strain to the respiratory system. The lungs' filtration reaches everywhere. The main functions of the respiratory system include the following:[8]

- providing oxygen at a cellular level
- eliminating carbon dioxide
- regulating blood's hydrogen-ion concentration (pH)
- forming speech sounds (phonation)
- defending against microbes
- influencing arterial concentrations of chemical messengers by removing some from pulmonary capillary blood and producing and adding others to this blood
- trapping and dissolving blood clots

A Bit of Anatomy

The heart distributes blood to every cell in the body. This makes oxygen and nutrients available throughout the entire system. Once

---
8   Widmaier, Eric P., Hershel Raff, and Kevin T. Strang. *Vander, Sherman & Luciano's Human Physiology: The Mechanisms of Body Function.* McGraw-Hill. 2003.

the cells and tissues of the body ingest the food from the blood the by-product is carbon dioxide. The veins that carry the blue blood back to the lungs do not rely on a pump; instead, they rely on muscle movement. The muscle movement initiates a pumping action, and the veins do the rest. In your veins, there are little one-way stop valves throughout. They are spread a proper distance apart from each other and only allow the venous flow to go toward the lungs.

When you move your muscles, you pump the venous flow upward. That is why exercise is so beneficial, even a daily walk around your block. This is also why you should not chronically cross your legs; this promotes varicose veins because your heart is getting the blood to your feet, but you are putting pressure on the vein and the blood can't escape properly. Over time, some of those one-way valves break because of the constant pressure; then that damage in turn plays off that pressure to the adjacent valves, and the domino effect begins. That is why cardiovascular exercise is highly recommended as part of a self-healing activity.

As humans in today's society, we spend most of our day shallow breathing. When we shallow breathe, we fail to properly deliver enough oxygen to the lower lobes of our lungs. Shallow breathing plays a part in anxiety. Take three deep breaths right now. That calm you feel going through your body is oxygen. That is how fast it is delivered. Take the example of public speaking, which most people fear more than death. You often see those souls struggling and gasping for air. Before walking in front of the crowd, this speaker was filled with fear and was not taking deep, relaxing breaths. Therefore starving themselves of oxygen; the synapses in their nervous system were firing off at an alarming rate. If the speaker had done a breathing exercise just before facing the crowd, a lot of the anxiety would have dissipated.

Look at this scenario from the way fibromyalgia proliferates in your body. The oxygen deprivation in the above example is also what is happening in your body, except it is high toxicity levels that are causing the biggest deprivation of oxygen to the blood, the result being the tissue. Once the saturation of toxins in the blood reaches these levels, the only action that delivers adequate reaction is opening up the floodgates. If ignored, the body's pH levels rise in acidity which also negates the body's natural healing process.[9] It's like you're drinking water with just a little bit of nuclear waste in it.

Let's now look at different levels of fibromyalgia. I say levels because if the disease is allowed to continue without resistance, then even after reversing fibromyalgia, certain established conditions may not be reversible.

With Danielle and other clients, once we reversed the fibro, the severe chronic arthritic conditions she had were not totally reversed. They were too far advanced. Another problem is that when a person experiences this condition, it puts his or her human chemistry in harmony with fibromyalgia and other cancers. Therefore, any evidence or trace evidence of inherited family disease will be accentuated during the life of this disease.

You have to wait until you reverse the fibromyalgia to find out what is left to battle. Diseases such as osteoporosis, arthritis, chronic depression, eating disorders, addictions, and intestinal problems are often the aftermath. Once fibromyalgia is reversed, then we see whether turning the water on over an extended period of time might help correct these conditions further. If diagnosed early and hydration is implemented immediately, that would be the best-case scenario for any individual coping with this disease.

---

9   Howell, Edward. *Enzyme Nutrition*. New York: Avery, 1985.

The bladder is the fourth toxin-dispensing organ. The bladder is the fastest organ in processing toxins through the body. As you know, fibromyalgia is severely affected when the fastest toxin-processing organ, the bladder, is malfunctioning. The bladder works on water; no water, no ATP, no work, and the garbage stays inside the body.

I continue to work with patients who have fibromyalgia, but they have to be ready for my boot-camp approach. When they are sick and tired of being sick and tired, then they are ready for me. You can find further info on my website: www.watertheuniversalhealer.com. Individuals with fibromyalgia need to unlock decades of toxin storage. If you never clean your body, nothing will work properly and you will break down.

Chapter Fourteen

Let It Go

*To be idle is a short road to death,
and to be diligent is a way of life.
Foolish people are idle; wise people are
diligent.*

—*Buddha*

The purpose of writing this book was to share a simple truth with you. I hope I was able to communicate that. I campaign for the fact that you are supposed to drink water. My purpose was to offer enough information for you to make an educated decision on the type of fluids you consume. I could offer much more information, but that would negate my original purpose in writing this book: keep it simple and factual.

My hope is that I have raised your awareness. I could continue discussing the importance of water in treating and preventing many more diseases. For example, I could also speak extensively on the effects of dehydration on diabetes, types I and II, but a single chapter would not be enough to explain it properly.

Masaru Emoto's brilliant work has brought to light the link between quantum physics and water. *The Message from Water* was an eye-opener. In it, Emoto proved that your simple intent could change the energy signal of a water cell. If water can react to strong words and feelings, can you imagine how it is interacting with our parched bodies? Imagine the impact your thoughts can have on your health.

My challenge to you is to stay properly hydrated with clean water for three months. See how much better you feel. Then, and only then, will you be able to experience the truth within you.

I was fortunate to have had a lot of great influences in writing this book. These men don't have a clue what they sparked within me. It goes without saying that what started me on this journey was *Your Body's Many Cries for Water*, written by Dr. Batmanghelidj, who opened my eyes to the facts of dehydration and the healing benefits of hydration. His book held my focus on this very important healing tool. If you wish to know more about the deeper science of human dehydration, read his book.

The Four Agreements and *The Fifth Agreement* by Don Miguel Ruiz were also instrumental to me. His teachings taught me to write by being impeccable with my words (the First Agreement). Ancient Toltec teachings might also be of some interest to you. His teachings help shape this message.

The Cure, written by Dr. Timothy Brantley, is a great book. Dr. Brantley was also inspired by *Your Body's Many Cries for Water*, and he goes into more detail regarding clustered water, which I just briefly touched on. *The Cure* will also explain nutrition in detail; if you read it, you will rethink any food you put in your mouth.

If you are interested in pursuing this line of thinking further, there is a plethora of information available. Hopefully, I have caught your interest, and you will let go of some of your old habits, and begin the journey with water, the universal healer.

Don't forget, dehydration is the disease.

*You've got one life, one life. Don't stop;
live it up. Go!*
—Hedley

Bibliography

Batmangheldij, F. *Your Body's Many Cries for Water.* Falls Church Global Health Solutions Inc., 2008.

Brantley, Timothy. *The Cure.* New Jersey: John Wiley & Sons, 2007.

Chopra, Deepak. *Perfect Health: The Complete Mind/Body Guide.* New York: Random House Digital Inc., 2001.

De Villiers, Marq. *Water: The Fate of Our Most Precious Resource.* Toronto: McClelland & Stewart, 2003.

Emoto, Masaru. *Messages from Water and the Universe.* Carlsbaad: Hay House, Inc., 2010.

Gascoigne, Stephen. *Manual of Conventional Medicine for Alternative Practitioners. Alternative Training.* 1995.

Gillespie, Gregg R. and Mary B. Johnson. *501 Recipes for a Low-Carb Life.* New York: Sterling Publishing Co. Inc., 2003.

Howell, Edward. *Enzyme Nutrition.* New York: Avery, 1985.

Moore, Keith L. *Clinically Oriented Anatomy.* 3rd Ed. Philadelphia: Williams & Wilkins, 1992.

Walker, N. W. *Fresh Vegetables and Fruit Juices.* Norwalk Press, 1970.

Walker, N. W. *The Natural Way to Vibrant Health.* Caroline House, Summertown TN, 1976.

Widmaier, Eric P., Hershel Raff, and Kevin T. Strang. *Vander, Sherman & Luciano's Human Physiology: The Mechanisms of Body Function.* McGraw-Hill, 2003.

Wilson, Lori. *Demystifying Medical Intuition.* Lori Wilson Educational Corp., 2005.